METAVERSE & MORTGAGES

A Journey to Global Mortgage Systems in a Digital World

For permissions and media inquiries:

Brigittejaward@infinitipremier.com

Personal Introduction

My journey began over three decades ago, in 1990, when I first entered the mortgage industry, and I was nicknamed "Ms. Finance." I cherished sitting with clients listening as they shared their hopes and goals while I worked to secure their dreams, brokering transactions between them and lenders.

During those early years, our work was defined by filing ledgers, using pens, handling paperwork, making phone calls, and engaging in face-to-face interactions. Pagers, pay phones, calculators, and fax machines were our lifelines, soon replaced by cell phones, computers, the internet, artificial intelligence, and WI-FI. A mortgage was a relationship, not just an application. I witnessed people's dreams, fears, and perseverance written into every form and closing document. Over time, their stories wove into my own and became part of me. They taught me that behind every file is a heartbeat and behind every person is a story.

As the years unfolded and I navigated different worlds, from traditional banks to global markets and digital frontiers, I watched technology transform that process into something faster, more efficient, and often more impersonal. The mortgage industry evolved from paper and ink to digital signatures, from local offices to global platforms, and from reality to virtual systems.

"*Metaverse & Mortgages*" was born from that question. This book reflects both my professional path and my personal conviction that housing finance is more than economics; it is humanity in numbers.

Every innovation we embrace, from blockchain verification to AI underwriting, carries an ethical dimension: how we treat people when money and technology intersect.

Throughout my career, I've been privileged to serve in diverse markets, including the United States, the Middle East, and other regions, each with its own unique rhythm and regulations.

Yet, all driven by a shared human need: the desire for stability, belonging, and opportunity. That universality inspired this exploration of the global mortgage system: its roots, risks, and reinventions.

In writing this book, I wanted to bridge the analytical with the emotional. The mortgage is a financial instrument. But, in truth, it measures human aspiration, a family's first key, an investor's goal, and government policies.

The future of mortgages, like the future of societies, will depend on how wisely we merge innovation with empathy.

The "metaverse" in this title is not an illusion; it is our new reality, the merging of digital and real life, where every data point represents a decision, and every decision shapes a person's future.

As we stand at this intersection, I hope readers will find in these chapters not only insights into markets but also a renewed perspective on what lies ahead.

To read *Metaverse & Mortgages* is to travel forward and see the evolution of trust, risk, and opportunity as both a human story and a technological saga. The metaverse is not fantasy; it is the next infrastructure of civilization. Within it, homes may float in code, but hearts will still seek stability.

This is my journey. And now, it becomes yours through the systems, histories, and practices that define how we live, build, and believe in the promise of home.

“*After all, everyone deserves a home.*”

—Brigitte Jaward-Hachem

Table Of Contents

Preface

To the pioneers who had the vision to imagine the future of digital finance long before it became mainstream: your courage to challenge convention laid the groundwork for the ideas presented in these pages. Your innovations and perseverance continue to redefine what is possible in property, equity, and technology.

Finally, to every reader who is inspired and curious, thank you for your interest and for engaging with our vision of change. The dialogue between past, present, and future is only meaningful when shared. And it is your engagement that brings the concepts within *Metaverse & Mortgages* to life.

From the first bank origins in Europe to the high-speed digital systems of today's markets, the mortgage has served as one of humanity's most enduring financial instruments. It binds individual aspiration, home, security, and stability to the broader structures of money, policy, and innovation.

This book illustrates how that connection has evolved. It examines how technological acceleration, global capital, and digital governance are transforming the ways people finance homes and how nations define wealth and inclusion.

Acknowledgments and Dedication

Projects of this depth rely on collective effort, research, and guidance. The contributions of professionals, researchers, and innovators have been helpful, shaping each chapter with deeper insights and data. This collaborative work sets the foundation for the future explored in these pages.

To my husband, children, family, and friends, thank you for your boundless love, patience, and faith that guided every word I wrote.

To my mother, thank you for all your prayers and blessings.

To my mentors and colleagues, whose wisdom, resilience, and vision continue to inspire my pursuit of excellence in every field I touch.

And to the dreamers, innovators, and global citizens shaping tomorrow's mortgage world, this book is for you.

Finally, to our readers: may this book enlighten your curiosity, inspire your courage, and spark your purpose as we all navigate the future landscape of finance and technology.

To the generations now and those to come, welcome to your new world!

The world of "*Metaverse & Mortgages.*"

Chapter 1:
Definition of a Mortgage

A A mortgage is a long-term loan secured by real estate, allowing buyers to trade time and trust for property ownership. The loan's purpose is to purchase real estate. The purchase applies to any property, including a house, an apartment, a commercial or residential building, and land. The property pledged as collateral for the loan, and a "lien" is recorded with the land department.

In a mortgage agreement, the borrower receives funds from a lender (usually a bank, financial institution, or private lender) to purchase the property and agrees to repay the amount over a fixed period initially agreed upon. It varies by country: in the USA, repayment periods and amortization schedules can be 10, 15, 20, or 30 years. However, some products could be interest-only, which extend to 40 years. Recently, a 50-year term mortgage was mentioned, but it is not yet approved.

Payments are regular monthly installments that include both the principal (the amount borrowed) and interest (the cost of borrowing). Both amounts determine the gross amount due.

If the borrower fails to make the required payments, the contract becomes due, and the lender has the right to reclaim the property through a legal process known as "foreclosure." In this process, the property is returned to the lender and sold at auction to recover the remaining debt.

For buyers unable to pay cash for a specific property, a mortgage allows individuals to own property by borrowing money upfront from a lender, paying the seller directly, and repaying the lender over time, while providing the lender with security through ownership rights until the loan is paid off.

Mortgages power the global economy. They create liquidity for banks, income for investors, and stability for households. Secondary markets transform individual loans into securities, feeding capital back into new lending. When functioning responsibly, this cycle powers mortgages in the global economy.

Future mortgage networks will integrate seamlessly into the individual's digital life.

Artificial intelligence (AI) advisors will interact and converse with borrowers, interpreting emotional tone and life goals before presenting financing options. These systems will learn from empathy as much as from economics: creating custom pathways that blend affordability with aspiration.

Real estate agents will evolve into professionals who curate physical and digital properties tailored to lifestyle, environmental preference, and virtual compatibility. The mortgage process will become less about qualification and more about identity: matching who you are with where you belong.

Types of Mortgages

1. Fixed-rate Mortgage (FRM): Interest rate remains constant; fixed monthly payments. It's an ideal product for long-term stability.

2. Adjustable-rate Mortgage (ARM): Interest adjusts periodically; it often starts with a lower rate but fluctuates based on margin and index. Lifetime caps apply to these products, but they become riskier when rates increase. There is another variable product called a "buy-down," which starts at a lower rate and gradually locks in at a higher rate over the fixed period of 2–30 years.

3. Alternative Financing, Interest-only Mortgage: Borrower pays only interest for an initial term; some of these products offer minimum payments that later create negative amortization of loans.

4. Conventional-conforming Mortgages, Jumbo-conforming/Non-conforming Mortgages, and Government-backed Loans: FHA, VA, similar

programs that expand access for first-time or lower-income borrowers.

5. Commercial and Construction Loans: Structured for income-producing or development properties; in most cases, they are short-term with balloon payments.

These types of mortgage products are available in the USA. Other countries have several similar programs.

Chapter 2:
The Players in a Mortgage Transaction

1. Borrower (Buyer, Individual or Entity)

The borrower is an individual or entity seeking to purchase real property or to obtain a mortgage loan to finance the purchase. In this process, a mortgage application and financial documentation—such as income and credit history—are submitted, and a down payment is required.

After all, the process is approved, a closing occurs. The borrower is responsible for making ongoing mortgage payments, paying property taxes, and maintaining insurance.

2. Lender (Bank, Financial Institution, or Private Lender)

The lender provides funding for the property. It can be a bank, a financial institution, a mortgage company, or a private lender. The lender evaluates the borrower's financial situation to determine whether the borrower qualifies for a loan.

After reviewing all documents and obtaining final approval, the closing process occurs, funds are disbursed, and the lender places a lien on the property. It serves as collateral until the mortgage is paid in full.

3. Mortgage Loan Officers and Mortgage Brokers

A mortgage loan officer is a licensed officer sponsored by a licensed mortgage broker; they act as a middleman coordinating between the borrower and multiple lenders. The borrower's application is reviewed and analyzed, packaged for placement with a matched lender.

Based on verified information submitted, the lender reviews and approves the loan in accordance with guidelines. The file closes, and funds are disbursed to all parties through a title or escrow agent.

Loan officers and brokers earn a fee or commission from either the lender or the borrower based on an initial agreement. But do not actually lend money themselves.

4. Real Estate Agent or Real Estate Broker

The real estate agent represents either the buyer, the seller, or both parties in a property transaction. They are licensed and sponsored by real estate broker companies.

They assist in locating properties, negotiating offers, and coordinating the sale process. While not directly involved in the loan, their role is essential in purchase and financing transactions.

5. Appraiser

An appraiser is a licensed real estate valuator assigned by the lender. Their task is to provide an independent evaluation and analysis of the property's market value, based on comparable and other relevant factors.

The lender uses this appraisal to ensure the property is valued at market rate and in proportion to the loan amount. This inspection report protects both the lender and the borrower from overpaying for the property and manages the collateral risk value.

6. Title Company or Attorney

Title companies and some states' real estate attorneys process title work, review it, and ensure that the property's title is clear of legal issues or ownership disputes.

They also handle closing documents, verify ownership history and chain of title, and issue title insurance to protect both parties from future title claims.

7. Closing Agent or Escrow Officer

The closing agent manages the final step of the mortgage transaction by acting as the closing or settlement coordinator. They transact and process the signing of all loan and property documents, handle disbursements, and ensure the property ownership is legally transferred from the seller to the buyer.

Chapter 3:
The Invention of Digital Finance

The year was turning into the 21st century, and the mortgage office was humming. Stacks of files leaned against the walls like fragile towers of history.

Every transaction was a ritual: brokers scribbled figures with pens, couriers rushed contracts through traffic, and underwriters judged each loan with caution. Fax and copy machines were feeding documents, and approvals had unlimited conditions; the process dragged on for weeks. Copy packages filled the cabinets with files and stories.

Then came the quiet pulse of a dial-up tone, the whisper of a world about to change. Computers began to hum in back offices, linking lenders to servers and servers to something significant and new: the birth of digital finance.

At first, it was modest, an email replacing a fax, a spreadsheet instead of a calculator. Yet, beneath that surface, the very foundation of money was shifting; data replaced paper. Information itself became a new form of collateral.

Lenders realized they could trade, analyze, and replicate that information faster than any courier could move a file. A mortgage no longer lived in a folder; it became a stream of code moving through networks.

In the USA, automated underwriting systems appeared, followed by similar models emerging in London, Frankfurt, and Singapore. Decisions that once required a seasoned officer's judgment now passed through algorithms that knew nothing of faces or families, only ratios and probabilities.

Digitization created efficiency, but it also lost the human element. The lender who once met the borrower now reads numbers instead of names. The art of lending began to yield to the logic of code. Still, the appetite for speed proved irresistible.

When the 2008 crisis occurred, it seemed the digital dream had failed. The markets collapsed.

Regulators demanded transparency; investors demanded truth. Borrowers demanded convenience. Cloud computing was created, with servers replacing filing cabinets and encryption replacing locked drawers. Compliance officers turned into data guardians, and underwriting evolved into engineering.

By the 2010s, a new generation of innovators emerged, building digital portals that instantly verified applications—APIs connected to credit bureaus, tax authorities, and even land registries.

The process that took weeks was shortened to hours. A borrower can apply for a mortgage from the comfort of their own home, never touching a sheet of paper or meeting the lender.

Meanwhile, in nations where physical structures were growing, the digital system became liberating. In India, biometric identity brought millions into the financial fold.

In Kenya, a mobile system transformed cell phones into wallets, giving rise to micro-lending systems that financed homes in previously unfinanced properties in villages. The *metaverse* of finance was a reality.

But with every innovation came challenges. Algorithms learned biases that their creators did not intend to instill.

Governments responded with privacy laws and digital safeguards. But mortgages were still about people. Behind every digital signature lies a hope, and behind every automated approval: a story.

Loan officers evolved into advisers through the maze of algorithms, serving as interpreters between the needs of machines and humans.

The invention of digital finance was about evolution. It turned ledgers into networks and transactions into conversations between code and conscience.

The future of mortgages would no longer be written in ink, but on a new platform across screens, servers, and the cloud. And yet, at its heart, it would remain a promise between humans built on trust and enduring through time.

Chapter 4:

Global Mortgage Systems and the Evolution

In the past, a mortgage was a handshake between two people. From medieval pledges of property to the rise of modern building societies, the mortgage carried one idea across centuries: that ownership could be both a dream and a duty.

By the mid-20th century, this dream had taken on new forms worldwide. In the United States, the mortgage became a symbol of democracy and stability. In Europe, centuries-old banks funded home loans through covered bonds and savings cooperatives.

Across Asia, state housing agencies directed credit to millions seeking shelter. Each region progressed at its own pace until the late 20th century, when technology and policy began to intertwine.

The 1970s and 1980s marked the beginning of this era. The U.S. pioneered mortgage securitization, bundling thousands of home loans into tradable instruments. The logic was appealing, turning a local promise into a global product.

Europe, cautious yet curious, began experimenting with similar models, while Asia and Latin America liberalized markets that had been closed to foreign capital. Deregulation and digitalization were the new system going forward.

By the 1990s, financial borders became borderless. A pension fund in another country could own part of a mortgage in Dallas or invest in Australian bonds; a family in Nairobi could benefit indirectly from the liquidity created in New York.

Mortgages, once intimate agreements between neighbors, became strands in a vast global web of money and data. The world's housing systems were all synchronized.

In 2008, the system's heartbeat failed. The collapse began with waves of default in suburban America and ended in a worldwide echo. The same technology that had spread opportunity now transmitted panic.

The promise of diversification became the reality of contagion. Governments, regulators, and central banks rushed to rebuild confidence: not only in markets but in the very idea that finance could be fair.

Out of that turmoil, digital accountability was increasing. Global accounting standards emerged, land registries went online, and shared data protocols replaced paper trails.

Cloud computing enables banks and lenders in various countries to operate on a shared digital infrastructure. Mortgage platforms became multi-lingual, multi-jurisdictional, and almost borderless. Transparency was no longer optional: it was the law.

Emerging economies and countries found themselves in this new era. Freed from the burden of legacy systems, nations such as India, Kenya, and Brazil have built digital frameworks that connect lenders, borrowers, and governments in real-time.

Yet, integration brought new questions: "Who owned the data? Whose ethics governed the algorithms? Could affordability be measured the same way in Los Angeles as in London?" Regulators began collaborating across borders, drafting shared principles for privacy, disclosure, and sustainability.

Today, the mortgage stands at a crossroads between the local and the global. Capital flows wherever opportunity calls. Yet, accountability remains anchored in national laws and social trust.

Policymakers walk a delicate line: trying to preserve domestic stability without erecting digital walls. For lenders and investors, innovation must be exercised with empathy, and expansion with ethics.

The evolution of global mortgage systems is not just a story of technology or trade; it is a story of humanity seeking permanence at an age of speed. Its essence remains unchanged. It is still a promise written in hope, signed in trust, and repaid across generations.

Chapter 5:
The Age of the Algorithm

There was a time when a banker's judgment could shape a destiny. He would review a paper file and decide whether a borrower is qualified for a mortgage.

The process was guided by mere judgment and experience, and most of all, it was personal. Somewhere in the conversation between borrower and lender, a relationship was built. Then, gradually, the world changed. The paper file became a data set. The banker's judgement became a code. And the decision became instantaneous.

The rise of the algorithm began with a single, simple goal: to make lending faster, efficient, and more consistent. Mortgage applications once took weeks; now decisions could be made in seconds. Behind the scenes, algorithms began evaluating creditworthiness, verifying income, and assessing risk with remarkable precision.

At first, it seemed simple: online portals replaced fax machines. Borrowers uploaded documents, and lenders verified details in seconds. But soon, the machines stopped waiting for instructions, as the elimination of human judgment allowed the algorithm to take the lead.

The spreadsheet evolved into a neural network. And the mortgage broker's checklist turned into a system capable of learning, adapting, and predicting.

In those early years, artificial intelligence (AI) could assess creditworthiness faster than any human officer, spotting patterns deep within large datasets. It could measure income stability, spending habits, and even forecast default probability.

The experienced loan officer's intuition, which could see beyond the numbers, was dismissed. Decisions once made by conversation became numerical and delivered instantly. For the first time, human judgment was replaced by data.

The industry embraced this efficiency with optimism. Entire loan files could move from application to approval without a single human hand. Automated valuation models replaced neighborhood appraisers, estimating home prices from historical records and data.

But behind all this, a question arose: “Could a machine understand human demands? Could an algorithm recognize determination, fear, or trust?” The models did not understand human behavior, the excitement of a first-time home buyer, or the pride of a family purchasing a home. They understood only data, with no touch of the human element.

Then came the regulators asking difficult questions. They discovered that the same algorithms designed to eliminate bias had sometimes learned to reinforce it.

Historic inequities had seeped into digital systems, invisibly guiding decisions. Whole communities once redlined by human prejudice now faced digital exclusion under the mask of neutrality.

Governments responded with a new language of ethics, fair lending algorithms, and data accountability. Finance was learning that technology still required morality. The algorithm could calculate risk, but it could not comprehend right and wrong.

By the late 2010s, the mortgage industry had become a vast ecosystem of automation. Robots handled servicing tasks, predictive software forecasted defaults, and investors traded algorithmically generated mortgage-backed securities. What had once been a relationship was now a network of decisions and profitability, but it lacked humanity.

Today, loan officers act as interpreters of code, translating algorithmic verdicts into understandable terms for humans. Borrowers still needed reassurance, still sought empathy. The machine could approve the loan, but only one person could explain the process to the applicant.

The age of the algorithm was not the end of the human element; it was its evolution. It taught the industry that speed means nothing without fairness, and precision is empty without compassion.

Tomorrow's algorithms will not only process information but also *sense* it. Integrated with emotional AI and behavioral analysis, they will distinguish between financial risk and human crisis. When a borrower falters, the system will adjust terms, guided by predictive empathy rather than punitive reaction.

Regulatory AI will serve as a global conscience monitoring system for real-time ethical compliance. Bias detection will become as fundamental as fraud prevention. Society will judge institutions not by profit, but by their digital morality.

The future of mortgage finance will always belong to innovation, but with the touch of humanity.

Chapter 6:
The Recovery After the Crisis

In 2008, the dark ages of the financial world: a moment when the world learned that confidence, once lost, can bring nations to ruin. Mortgage markets, once symbols of stability, became the fault lines of collapse.

Families lost homes, banks lost faith, and governments faced the sobering truth that the system built to create security had instead magnified fragility.

In the days that followed the crash, silence replaced certainty. Screens once glowing with profit turned red with panic. The air in trading floors grew heavy, thick with disbelief. The financial world collapsed.

Central banks intervened first, pumping liquidity to unthinkably high levels. Interest rates fell to near zero. Governments stepped in to purchase mortgage-backed securities, guaranteeing deposits, and rewriting the rules of survival. It was not merely a rescue; it was a reinvention.

After the financial storm, a new mindset began to emerge. The mortgage industry, chastened and cautious, turned to technology as both shield and solution. The same digital tools that once fueled overconfidence became instruments of reform: an instrument of survival.

Regulators demanded data records and trails, rather than assumptions. Investors insisted on real-time reporting. Borrowers demanded clarity. The industry responded by transforming paper into pixels and processes into protocols. Manual document verification gave way to digital audits.

Signatures became secure e-signatures; compliance officers became architects of ethics.

In the early 2010s, recovery was crucial. The mortgage office, once measured by volume, began to prize resilience. Lenders spoke of "sustainability" not as a slogan but as a survival code. Risk management became the innovation; ethics became a business model.

Technology stepped into this new world with a quiet determination. Data analytics platforms allowed institutions to stress-test their portfolios under countless scenarios.

Artificial intelligence (AI) began identifying early signs of default before they spiraled into losses. Cloud computing has made collaboration across continents effortless, enabling regulators, investors, and lenders to access shared information more easily and efficiently.

Even as systems modernized, the human cost lingered. Millions who lost homes carried stories that no algorithm could erase. A new generation of professionals has entered the mortgage industry: determined to strike a balance between precision and compassion. They sought to rebuild, innovate, and bring back the promise of mortgages.

By the mid-2010s, regulations and accountability tightened, and mortgage lenders once again became stewards of trust. Capital strength replaced speculative appetite, and long-term value replaced short-term gain.

Technology, regulation, and humanity became partners. The industry discovered that innovation without accountability was risky, and oversight without empathy was stagnant.

The very future of finance would depend on blending all these elements; it was the dawn of a new order. The world has learned that systems fail when they forget the people they serve. The next chapter of mortgage finance would be about endurance and survival.

Chapter 7:
The New Financial Foundation

In the decade following the financial crisis, the global economic system underwent significant reconstruction. Governments demanded transparency, investors demanded efficiency, and consumers demanded accessibility.

The result was the restructuring of payment systems, mortgage servicing platforms, clearing houses, and data networks. Legacy mainframes gave way to cloud-based infrastructure capable of processing and remote integration.

Financial institutions recognized that infrastructure was a crucial competitive advantage. Banks began treating their technology stacks as strategic assets, measuring latency and cybersecurity alongside revenue. Mortgage firms and lenders digitized everything from loan origination to escrow management.

During that time, these technologies defined the new reality. Cloud computing enabled lenders to store and process information on a scale, thereby reducing costs.

Application programming interfaces (APIs) connect disparate systems, allowing credit bureaus, insurers, and regulators to share verified data. And, in its early stages, blockchain introduced the idea of a shared ledger that could authenticate ownership and transaction history with mathematical certainty.

Digitization brought new challenges. As lenders moved online, cybercrime expanded. Data breaches, ransomware, and identity theft threatened not just individual borrowers but systemic stability.

Regulators began classifying cybersecurity as a component of serious supervision. Institutions implemented encryption and biometric authentication.

Resilience was necessary. The post-crisis world learned that continuity of financial operations was imminent. Cloud servers, disaster recovery sites, and regulatory stress tests now rule every central mortgage platform. The COVID-19 pandemic, which forced global remote work, validated these investments; the digital rails held firm.

Today, financial infrastructure operates remotely off-site. Millions of borrowers sign in to an app to check a balance or submit a document, unaware that the data travels through hundreds of interconnected nodes governed by international standards. This invisible network is the nucleus of the global mortgage economy.

A future borrower in Japan may use AI to compare global lending options. At the same time, a family in Nigeria will secure a home through a solar-powered micro-grid mortgage verified in the cloud. East and West will no longer represent divisions, but dimensions of the same evolving financial consciousness.

Innovation is accelerating, and digital and other forms of payments are underway; one example is Stablecoin.

Stablecoins, which are digital currencies backed by stable assets, are simplifying and accelerating mortgage transactions. They enable instant, low-cost payments, especially for international deals, eliminating traditional banking delays. Through smart contracts, stablecoins automate tasks such as escrow release and payment confirmation, reducing risk and speeding up closings. For lenders and investors, they support tokenized mortgage products and faster distributions. Overall, stablecoins help create a more efficient, transparent, and globally accessible mortgage system.

The next frontier is cultural integration. Technology will illuminate differences. Mortgage systems will adapt to local values while speaking a global language of transparency. Islamic finance will coexist with algorithmic lending; African mobile platforms will merge with Western securitization.

Chapter 8:
Globalization of Mortgage Lending

In the 2020s, mortgages became a global asset class: pension funds in Toronto financed homes in Seoul, sovereign wealth funds in Abu Dhabi held stakes in European securitizations, and Asian insurers invested in U.S. agency bonds. The search for yield and diversification was a global phenomenon.

Digital platforms accelerated this globalization. Marketplace lenders matched investors and borrowers across jurisdictions, while international payment rails allowed capital to move seamlessly between currencies. Yet, this freedom introduced new dependencies: currency volatility, regulatory divergence, and exposure to global shocks.

For developing nations, access to global liquidity has significantly transformed local housing finance. In Latin America, covered-bond frameworks mirrored those in Europe; in Africa, regional mortgage refinance companies aggregated loans for securitization purposes.

However, globalization also demands governance. Transparency in underwriting, data quality, and servicing has become a prerequisite for attracting capital.

Digital reporting portals have replaced spreadsheets. And regulators in Nairobi, Bogotá, and Manila have learned to speak the same data language as investors in London or New York.

The integration of global lending meant that financial contagion could now travel through code as easily as through capital. To counter this, the Financial Stability Board (FSB) and the IMF have coordinated macroprudential toolkits: policies that enable countries to regulate or stimulate housing credit without isolating themselves. Central banks began sharing analytics on mortgage flows, and supranational organizations established crisis-response protocols.

The globalization of lending is ultimately a paradox; it unites markets while challenging sovereignty. Capital seeks the most efficient path, but regulation remains territorial. The task of the coming decade is to create a global common of credit, a framework in which information flows freely and local borrowers benefit from global liquidity.

Chapter 9:
The USA Mortgage System

The United States remains the nucleus of the world's mortgage ecosystem. With more than $12 trillion in outstanding debt, it represents the deepest and most liquid housing finance market globally. Its dual-agency structure, with Fannie Mae and Freddie Mac, anchors investor confidence by guaranteeing standardized securities backed by home loans.

The U.S. mortgage is both a domestic necessity and a global benchmark. Yields on agency bonds influence capital flows from Tokyo to Frankfurt. American underwriting innovations, from fixed-rate amortization to digital eClosings, set templates that other nations adopt or adapt.

Post-2008 reforms tightened oversight and transformed the industry's culture. New regulations established stricter documentation requirements. Enforced consumer protections and new capital rules constrained excessive leverage.

At the same time, technology reshaped operations: with eMortgages, hybrid appraisals, and instant income verification, the approval process was reduced from weeks to days.

The market's resilience during the COVID-19 pandemic demonstrated how digitization and policy could coexist. Forbearance programs protected millions of homeowners, while remote-work systems kept the industry functional. Fintech lenders proliferated, serving niches overlooked by traditional banks, yet the agencies maintained systemic stability.

Despite modernization, affordability remains a persistent crisis in America. Rising home prices, stagnant wages, and constrained supply have widened inequality.

Policymakers are experimenting with shared-equity loans, down-payment assistance, and green-mortgage incentives to align housing policy with climate and social goals. Technology, now, enables targeted subsidies based on data analytics rather than broad subsidies that inflate prices.

The U.S. model illustrates that mortgage liquidity can drive national prosperity, but also that financial innovation must be accompanied by vigilant regulation as global markets digitalize.

Technological advancements, demographic shifts, and new financial regulations will drive the future of the U.S. mortgage market.

As digital platforms continue to transform the lending industry, the mortgage process has become faster, more transparent, and increasingly data-driven. Artificial intelligence (AI) and blockchain technology may redefine underwriting, appraisal, and title verification, reduce costs, and shorten approval timelines.

Looking further ahead, the mortgage market will also respond to generational shifts in housing demand and economic priorities. Millennials and Gen Z buyers, who value flexibility and digital convenience, are already influencing the design of loans and property trends.

Climate resilience and green financing are likely to become integral to mortgage products, as lenders account for the long-term risks associated with extreme weather and prioritize energy efficiency, interest-rate volatility, global capital flows, and changing work patterns.

Chapter 10:
The UK and European Mortgage Systems

Europe's mortgage system is defined by national traditions bound by shared regulatory threads. The eurozone's single monetary policy coexists with diverse housing systems, Germany's long-term fixed-rate loans, France's state-supported savings plans, Spain's variable-rate culture, and the UK's competitive broker network.

The 2008 crisis prompted continental reform. The mortgage credit regulators encouraged consumer protection. Covered bonds, centuries-old instruments backed by high-quality assets, proved remarkably resilient: supplying the industry with a stabilizing core while securitization markets were rebuilt.

Brexit reshaped the geography of finance but not its essence. London retained its role as a global hub, while Dublin, Frankfurt, and Amsterdam absorbed portions of neighboring businesses. UK lenders are focused on agility, embracing open banking frameworks and digital identity verification.

Europe leads the world in integrating sustainability into mortgage policy. Under the European Green Deal, banks assess the energy performance of buildings as part of their credit evaluation.

Green mortgages offer preferential rates to borrowers who upgrade insulation or adopt renewable energy systems. Data-sharing platforms now track carbon footprints of collateral: linking environmental metrics with financial performance.

The European experiment shows that cooperation need not erase diversity. By balancing national identity with continental objectives, Europe demonstrates that finance can be both integrated and ethical. Its experience offers a template for regions seeking unity without uniformity.

Chapter 11:

Asian, Middle Eastern, and African Mortgage Systems

Across the developing world, the mortgage is more than a financial product; it is a tool. Urbanization, rising incomes, and demographic growth have made housing finance a central component of economic policy in Asia, the Middle East, and Africa. Governments view mortgage expansion as both an economic stimulus and a means of social stability.

In Asia, policy and innovation converge. China's massive housing-finance complex operates through state banks and urban-development funds, striking a balance between growth and political oversight. India's Pradhan Mantri Awas Yojana (PMAY) program mobilized millions of small borrowers using digital identity verification under "Aadhaar."

Southeast Asia's fintech lenders, particularly in Indonesia and Vietnam, have developed fully mobile underwriting systems that reach borrowers with no prior banking history.

In the Middle East, diversification away from oil has made real estate finance central to national visions. The UAE's blockchain-enabled land registries, Saudi Arabia's long-term fixed-rate mortgages, and Qatar's sovereign-backed housing funds illustrate a hybrid model that combines traditional capital with modern technology. Islamic finance principles, emphasizing asset backing and shared risk, encourage prudence while promoting inclusion.

Africa faces distinct challenges: low incomes, limited collateral, and high construction costs. Yet, innovation thrives. Kenya's micro-mortgage apps, Nigeria's blockchain title registries, and Rwanda's public-private partnerships with regional refinance facilities are redefining accessibility in the region. Regional development banks provide credit lines and guarantee access to other sources of funding.

The unifying feature across these regions is adaptability. With fewer legacy systems, they are free to design mortgage markets tailored to their needs.

Digital identity, mobile payments, and decentralized data systems allow them to skip intermediary stages of modernization. The result is not imitation, but reinvention: a new model of inclusion built for the digital age.

Chapter 12:

Canadian, Australian, and South American Mortgage Systems

Canada and Australia share conservative lending cultures that value stability over growth. Strict underwriting and recourse lending helped keep default rates low. Nevertheless, soaring property prices in cities like Toronto and Sydney revealed affordability limits even within well-regulated frameworks.

Both countries integrated technology early. Digital mortgage portals, open banking connectivity, and instant verification systems make their markets more efficient and transparent.

Regulators balance innovation with caution; Canada's system requirement ensures borrowers can withstand higher interest rates. And Australia gives borrowers control over their financial information.

In South America, structural reforms are bringing long-term mortgage markets to maturity. Brazil's system anchors state policy, while private-sector securitizations fund middle-class housing.

Chile's mortgage system remains a global reference for preserving real value. Digitalization is accelerating in Mexico and Colombia, where fintech partnerships bridge informal employment gaps.

Currency volatility and inflation continue to challenge affordability, yet technology offers potential mitigation. Real-time pricing, blockchain registries, and regional credit bureaus improve transparency and investor confidence. Continental cooperation through development banks fosters shared infrastructure for data and payments.

Across these nations, the guiding principle remains one of balance, innovation with prudence, and accessibility with stability. They illustrate that modernization need not come at the expense of resilience.

Chapter 13:

Reverse Mortgages: A Comparison of USA and Global Systems Versus Forward Mortgages

If humanity lives longer, wealth must evolve with time. Reverse mortgages are becoming a cornerstone of aging societies. They allow seniors to unlock the value of their homes without leaving them: converting walls into wealth and comfort into continuity.

In the United States, reverse mortgages offer monthly income or lines of credit secured by decades of property ownership. Europe emphasizes consumer protection and education. Australia and Canada champion regulated, government-insured models.

In Asia—where family remains the primary safety net—reverse mortgages are redefining intergenerational support, transforming inheritance into a means of living with dignity.

The future of longevity finance will intertwine biology and blockchain. Smart contracts linked to health data will dynamically adjust repayment terms, easing burdens during illness and rewarding longevity with lower rates.

Retirement homes may be tokenized, allowing investors to fund aging communities through micro-mortgage pools that yield both profit and compassion.

As populations age, reverse mortgages—which convert home equity into income—have become essential to retirement planning. Programs such as the U.S.'s Home Equity Conversion Mortgage (HECM) and the UK's lifetime-loan schemes demonstrate how financial engineering can serve social policy. In Japan and South Korea, the government provides support to elderly homeowners who lack pension coverage.

Forward mortgages, also known as new written mortgages, conversely, remain the foundation of wealth creation for younger generations. The two forms represent opposite flows of capital: one monetizes accumulated equity, while the other invests in future ownership. Together, they form a continuum that links life cycles, savings, and intergenerational policy.

Transparency and consumer protection are crucial. Reverse mortgages require careful counseling to prevent exploitation; forward loans demand clear risk disclosure.

Policymakers are increasingly viewing housing finance as part of a holistic financial-wellness framework, integrating pensions, insurance, and healthcare into a single ecosystem.

Reverse Mortgages (U.S. and Global Perspectives)

Reverse mortgages are a financial tool that allows older homeowners to access their home equity without selling their homes or making monthly mortgage payments.

In the United States, some reverse mortgage products are federally insured, providing seniors with a safety net and a predictable structure to access funds as lump sums, monthly payments, or lines of credit. The borrower retains home ownership. But the loan balance grows over time as interest and fees accumulate.

Over time, reverse mortgages have evolved from niche products into mainstream financial planning tools for retirees seeking to supplement their income, cover healthcare costs, or enhance their quality of life.

Yet still, challenges remain, particularly regarding education, fees, and potential misunderstandings about long-term obligations. Regulators continue to refine oversight to ensure borrowers understand their rights and responsibilities.

Reverse Mortgages in Other Countries

While reverse mortgages have become a well-known financial instrument in the United States, similar models have also emerged in various parts of the world: each shaped by distinct legal, cultural, and economic frameworks.

Across Canada, the United Kingdom, Australia, Germany, and beyond, the concept of unlocking home equity in retirement has gained traction—albeit with varying names, structures, and levels of maturity.

In Canada, reverse mortgages are available to homeowners aged 55 and older, allowing them to borrow up to about 55% of their home's value. The Canadian model operates under consumer-protection guidelines like those in the U.S., but with less government insurance backing.

Homeowners are not required to make monthly payments; the loan is due upon the borrower's death or upon vacating the property. The Canadian approach reflects a growing market driven by aging homeowners seeking flexibility while maintaining ownership of their homes.

In the United Kingdom, reverse mortgages are commonly referred to as equity release mortgages. This market has expanded rapidly among homeowners aged 55 and over.

Unlike the U.S. model, which relies heavily on federal insurance programs, the UK system is shaped by private-sector lenders and industry standards established by the Equity Release Council (ERC).

Similar equity-release concepts are gaining traction across continental Europe, where demographic trends and housing wealth indicate growing long-term potential, despite slow adoption.

Australia has also developed a regulated reverse mortgage framework, typically available to homeowners aged 60 or older. Borrowing limits depend on age, property value, and location. And products are governed under national consumer credit laws.

Over the last decade, tighter oversight and clearer rules have helped strike a balance between borrower protection and lender participation, allowing retirees to access their home wealth while retaining occupancy rights.

Reverse mortgages will evolve into *a special loan* instrument that fosters empathy, sustaining independence rather than surrendering it. In an age obsessed with youth, they will remind us that stability, not novelty, is the ultimate innovation.

Chapter 14:

A Unique Product: About Islamic Financing in the Global Market

What is Islamic financing, and how does it differ?

Islamic financing is designed to operate without charging interest, known as "Riba", which is strictly forbidden under Islamic religious laws and beliefs.

Islamic home finance aligns housing credit with Shariah laws and policies that prohibit riba (interest), gharar (uncertainty), and maysir (speculation).

Instead of charging or paying interest, Islamic structures use asset-backed transactions such as sale, lease, buy-back, or partnership. This model ensures financing remains tied to the real economy while maintaining fairness and transparency.

These are core principles:

1. Asset-backing: Financing tangible assets, specifically real property.

2. Risk-sharing: Both the lender/bank and the customer share ownership or risk in the asset.
3. Transparency: Disclosure of Prices, profit margins, and responsibilities.
4. No Interest (Riba): Profit is permissible from trade, buy back, or lease, not from the time-value of money. Mainly consists of home-finance structures.
5. Murābaḥah (Cost-plus sale): The buyer purchases the property back from the bank or lender at a markup agreed upon, payable over time.
6. Ijārah muntahiyah bittamlīk (Lease-to-own): The bank leases the property and transfers ownership gradually or at lease-end.
7. Mushārakah mutanāqiṣah (Diminishing Partnership): The buyer and bank co-own the home. The buyer gradually purchases the bank's share.
8. Istisnāʿ (Construction Financing): Used for building projects with progressive ongoing payments.
9. Sukūk (Islamic Bonds): Used to fund pools of Shariah-compliant housing finance.
10. Equity Release and Reverse Models: While conventional reverse mortgages rely on interest and compounding, Islamic finance achieves

similar goals through sale-and-leaseback, decreasing-maturity, or agency-based structures in which the institution earns rent, not interest. These methods maintain Shariah compliance while supporting retirement liquidity and flexibility.

11. Risk, Pricing, and Compliance: Since Islamic financing avoids interest, many products benchmark profit rates against market indicators to ensure competitiveness. Shariah supervisory boards ensure compliance, while risk management covers title, insurance (takāful), early settlements, and defaults.
12. Global Practice Malaysia and GCC: Advanced markets with clear regulatory frameworks and sukūk funding.
13. United Kingdom: Islamic home-purchase plans operate under Financial Conduct Authority (FCA) oversight with tax parity.
14. North America: A growing market through cooperative and fintech partnerships.
15. Emerging markets: Governments integrate Shariah housing programs with affordable-housing policies.
16. Digital Transformation: Open banking and smart contracts enhance transparency in Islamic financing. Blockchain and tokenization can

automate equity-share tracking and ensure compliance with Shariah rules.

17. Consumer Protection and Disclosure Borrowers: Information about cost comparisons with conventional loans. Late-payment penalties cannot generate profit; they are donated to charity.
18. Sustainability and Policy Integration: Islamic finance naturally supports environmental and social goals because of its real-asset basis. Green sukūk and energy-efficient home financing combine ethical and ecological principles.
19. Outlook: Future systems will adopt standardization of documentation, secondary-market liquidity through sukūk, and the broader adoption of fintech platforms to serve both global Muslim and ethical investors.

Chapter 15:

Innovation, Regulation, and the Future

In today's world, innovation and regulations are partners. Governments now create regulatory platforms that allow fintechs to test products under inspection.

The UK pioneered the model, with Singapore, Abu Dhabi, and the U.S. states following suit. The approach encourages experimentation while maintaining oversight and control.

Artificial intelligence (AI), open data, and ledgers demand governance. The future of regulation is based on measuring fairness, transparency, and systemic impact, rather than relying on technology.

Global coordination and collaboration are increasing through the Financial Stability Board (FSB), the IMF, and regional bodies to prevent regulatory arbitrage.

Even in a time of automation, expertise remains essential. Financial literacy, ethical leadership, and diversity in data science ensure that digital transformation benefits society. Regulation is no longer reactive; it is co-creative, shaping markets in real-time.

Chapter 16:

Survival in Global Mortgage Finance

Climate risk has become a central element of financial stability. Hurricanes, floods, and wildfires can destroy collateral overnight, threatening lenders and communities alike. Central banks now classify climate exposure as systemic risk.

Green mortgages incentivize energy-efficient construction and retrofits through lower rates. Sustainability also encompasses inclusion and ethics.

Fair access to credit and lending, and transparent corporate governance, are now integral to resilience. Financial institutions recognize that inequality undermines repayment capacity just as surely as inflation or unemployment.

In this sense, sustainability is not philanthropy: it is risk management. A sustainable mortgage system protects both people and portfolios.

Chapter 17:
The Future of Global Mortgage Systems

The convergence of technology, policy, and capital has created a global financial web. Mortgages now exist as data objects that are tokenized, traceable, and tradable.

The "metaverse" of finance is not virtual reality, but digital interoperability: systems that communicate seamlessly across borders and platforms.

Smart contracts automate disbursements, digital twins simulate housing developments before construction, and fractional ownership allows households to access capital without selling the entire property. The potential for efficiency is immense, but so is the responsibility to ensure equity and privacy.

The ultimate purpose of mortgage innovation remains unchanged: to provide individuals and families with a secure, stable, and dignified home.

Future success will depend on aligning algorithms with human values. The next generation of mortgage systems will be judged not by speed or yield but by inclusion, resilience, and integrity.

Chapter 18: The Future of Global Real Estate

Unprecedented technological, environmental, and social transformations are shaping the future of global real estate. Urbanization continues to accelerate, but the form of cities is changing vertically, digitally, and increasingly sustainable.

Smart infrastructure—powered by artificial intelligence and the Internet of Things (IoT)—is redefining how buildings operate, manage energy, and serve their occupants. Real estate assets are evolving from static investments into dynamic data-driven platforms.

Emerging markets are becoming the new frontier for growth as rising middle classes in Asia, Africa, and Latin America create surging demand for housing and commercial property. Global capital flows are diversified: moving beyond traditional hubs like

New York, London, and Hong Kong are among cities such as Singapore, Dubai, Nairobi, and São

Paulo. Technology is lowering barriers to cross-border ownership: enabling fractional real estate investment and tokenized assets that democratize access to property management.

At the same time, the climate crisis is forcing a fundamental rethinking of where and how we build. Rising sea levels, extreme weather, and resource scarcity are prompting governments and developers to prioritize green architecture, renewable energy, and circular design principles.

Cities of the future will rely on data analytics to forecast environmental risks, while adaptive reuse of existing structures will replace wasteful demolition.

Demographic shifts and new work patterns redefine global property demand. Remote work has reduced dependence on centralized offices and created a hybrid geography of work-live spaces, co-living models, and flexible micro-developments. The next generation of buyers seeks mobility, inclusiveness, and a sense of meaning in ownership.

Ultimately, the global real estate landscape is entering a new era where technology, sustainability, and human experience converge. The most successful cities and developers will be those that blend innovation with need, thereby designing environments that are intelligent, inclusive, and resilient.

In this interconnected future, real estate will no longer be simply a foundation of wealth, but a reflection of global progress and shared responsibility.

Epilogue:

The Human Element, Generation Rewriting the Rules, and What Tomorrow Brings

The story of the mortgage is, in many ways, the story of civilization: reflecting the evolution of trust, property, and promise.

As we enter an era shaped by algorithms, automation, and artificial intelligence (AI), the mortgage remains an enduring symbol of continuity. It has evolved from papers sealed in ink to blockchain-verified smart contracts, yet its essence remains unchanged: a human promise measured by numbers but fulfilled by dreams.

In the 20th century, the mortgage was an act of faith between two parties. The lender relied on paper, proof, and proximity; the borrower relied on trust. Today, that faith is migrating into digital systems, encrypted and algorithmically enforced.

The numbers tell their own story. Globally, outstanding residential mortgage debt now exceeds USD $59 trillion, representing over half of all consumer credit.

In advanced economies, such as the United States, mortgages account for over 65% of household liabilities. In developing regions, the share is smaller but growing exponentially as technology lowers barriers to access.

Digital platforms have redefined the application process. What once took weeks of manual verification now takes minutes. Algorithms assess creditworthiness by analyzing income data, payment history, and even behavioral analytics. Blockchain networks securely record property titles, reducing fraud and simplifying ownership transfers.

These innovations have made the mortgage system faster and more secure, but they have also introduced new complexities. Data bias, cybersecurity risk, and the digital divide now stand as the modern equivalents of past credit crises.

In the emerging digital mortgage economy, automation has become the driving force behind growth. AI-driven underwriting tools accurately predict borrower risk.

Fintech companies utilize machine learning to provide dynamic interest rates tailored to individual profiles. Cloud-based data storage enables cross-border verification in seconds, supporting international investors and remote transactions.

Global lenders have already integrated AI underwriting and blockchain-based collateral management, while governments in the United Arab Emirates and Singapore are building national digital land registries. Meanwhile, in Africa and South Asia, mobile lending applications are giving first-time homeowners access to formal credit markets.

The true challenge of the digital mortgage era is not innovation but building inclusive systems that remember that a mortgage is not just statistical data, but a human story.

The word "metaverse" once belonged to science fiction. Today, it represents a new financial frontier where physical and digital assets coexist, and ownership extends beyond geographical boundaries.

Virtual real estate, tokenized property, and decentralized finance (DeFi) platforms are rewriting the boundaries of value. Investors now buy digital land plots, finance construction in the virtual world, and secure loans backed by non-fungible tokens (NFTs).

Many institutions have entered the metaverse through projects for digital mortgages and blockchain-based collateral systems.

This merging of virtual and real estate reveals the next chapter in human finance: one where data is the deed, trust is encrypted, and location is secondary to identity.

Technology continues to blur boundaries between nations, industries, and realities. Yet, every algorithm still reflects human intent. The challenge ahead is not only efficiency but empathy: ensuring that the systems we build serve the people who depend on them.

In the age of artificial intelligence (AI) and the digital era, the mortgage remains one of the last places where data and emotion intersect. It is the intersection of dreams and due diligence, of trust and verification.

Behind every blockchain ledger and algorithmic model is a person, a borrower trying to build a home, a lender assessing risks, a policymaker shaping fairness. The future will not be defined solely by the intelligence of our systems, but by the integrity of our choices.

Automation may streamline the process, but only compassion can sustain progress. As finance becomes more predictive, we must ensure it remains inclusive; as technology becomes more autonomous, it must remain accountable.

The metaverse of mortgages is not about escaping reality; it is about enhancing it. It offers a chance to design systems that respect both the efficiency of data and the dignity of human life.

Yield curves or tokenized assets will not measure the ultimate success of this evolution, but rather whether families can still feel safe turning a key in their own door.

The next generation of borrowers is rewriting the rules, just as every generation before it has. However, this generation, the digital natives, is born in the shadow of crisis and codes rewriting the rules altogether.

The traditional dream of "buying a home and settling down" has been reframed. Today's borrower is not defined by geography or stability but by connectivity and adaptability. They move between cities, countries, even virtual spaces, carrying their credit and earning powers in their pockets.

The next generation will not inherit traditional debt; they will inherit intelligent equity. Every transaction, every mortgage, every asset will be part of a living portfolio that grows and adapts in sync with the individual.

What tomorrow brings is a borderless future, but not an equal one. Those excluded from the digital economy—the disconnected, the unbanked, the undocumented—risk being written out of financial relevance altogether.

Our task is to ensure that every advancement brings us closer to the human element. An element that can never be replaced.

Ultimately, the mortgage is not merely a financial contract; it is a cultural covenant. It measures the rhythm of civilization, our need for security, identity, and continuity. The metaverse may host our next frontier, but the essence of home remains eternal.

Tomorrow's finance will merge wealth and wisdom with a purpose, as humanity learns to code empathy into economics, ushering in the age of the metaverse and mortgages.

—Brigitte Jaward-Hachem

Research disclosure

All references listed in this work are publicly available institutional or regulatory publications used solely for factual validation and educational context.

No proprietary, confidential, or copyrighted materials have been reproduced. Data and insights are presented in good faith to enhance understanding of global mortgage systems and digital finance.

Declaration of low use of generative AI and AI-assisted technologies for correction and clarity.

References and sources

The following references represent primary institutional, regulatory, and economic sources consulted for factual and comparative insights throughout this work. Inline citations are intentionally omitted to preserve narrative clarity and readability.

- Australian Prudential Regulation Authority (APRA). Housing and Lending Guidelines.

- Bank of England and Financial Conduct Authority (FCA). Mortgage Credit Regulations and Open-Banking Standards.

- Canada Mortgage and Housing Corporation (CMHC). Residential Mortgage Market Analysis.

- Central Bank of the United Arab Emirates (CBUAE). Mortgage and Blockchain Land Registry Regulations.

- Consumer Financial Protection Bureau (CFPB). Fair Lending and Mortgage Disclosure Rules.

- European Central Bank (ECB). European Mortgage Markets and Covered Bond Reports.

- Federal Housing Administration (FHA). U.S. Mortgage Programs and Homeownership Data.

- Federal National Mortgage Association (Fannie Mae) and Federal Home Loan Mortgage Corporation (Freddie Mac). Mortgage-Backed Securities and Secondary Market Guidelines.

- Financial Stability Board (FSB). Macroprudential Policy and Mortgage Regulation Frameworks.

- International Monetary Fund (IMF). Global Financial Stability Reports and Housing Finance Data.

- Islamic Financial Services Board (IFSB). Guiding Principles for Shariah-Compliant Home Finance.

- Kenya Mortgage Refinance Company (KMRC). Digital Mortgage and Housing Finance Initiatives.

- Reserve Bank of India (RBI). Housing Credit and Digital Finance Frameworks.

- World Bank Group. Global Housing Finance Series and Affordable Housing Initiatives.

- World Economic Forum (WEF). Reports on Digital Finance, Blockchain, and the Future of Real Estate.

About the Author

BRigitte Jaward-Hachem is an entrepreneur and a pioneer in the mortgage industry, with over three decades of experience spanning mortgage lending, real estate sales, training, consulting, banking, and capital markets.

She is the founder and CEO of Infiniti One Corp., a U.S.-based, licensed mortgage-lending brokerage registered with the Nationwide Mortgage Licensing System (NMLS) (NMLS #318302). And the principal broker/owner of Premier One Group Inc., a real estate firm in the USA serving domestic and international clients.

Brigitte has academic and advanced professional credentials in higher education, finance, and real estate, including multiple licenses and certifications. Her career has spanned U.S. and global markets, including advisory work in the United Arab Emirates (UAE).

She is a leading industry figure in the United States, a champion of digital transformation, and a proponent of sustainable housing finance.

Brigitte knows "*The Art of the Mortgage,*" how to read beyond numbers, interpret borrower behavior, and structure solutions that align financial goals with human needs.

Her analytical insight bridges intuition and data, giving her a unique ability to assess risk, evaluate opportunity, and guide clients through complex financial landscapes with confidence, clarity, and efficiency.

She also serves on several boards related to business, real estate, and finance, as well as city and state commissions focused on the mortgage industry, finance authority, tax increment, and business development. This involvement contributes thought leadership to the evolving intersection of global finance, technology, and ethical governance.

Disclaimer

THis book is intended solely for general educational, informational, and discussion purposes. It does not provide legal, tax, investment, mortgage, accounting, or regulatory advice and should not be relied upon for professional advice.

Laws, regulations, product terms, and market conditions change frequently and vary by jurisdiction and geographic location; examples and references may not apply in your area or at the time of reading.

While reasonable efforts were made to ensure the accuracy and timeliness of the information presented, no representation or warranty, express or implied, is made as to its completeness or fitness for any purpose.

Mentions of specific geographic regions and mortgage systems, technologies, or financial products are for illustration only and do not constitute endorsement.

Any view expressed is that of the author in her personal capacity.

Made in the USA
Monee, IL
03 November 2026

67379250R00049